REEDS
CLOUD
HANDBOOK

OLIVER PERKINS

T0025740

REEDS
LONDON · OXFORD · NEW YORK · NEW DELHI · SYDNEY

REEDS
Bloomsbury Publishing Plc
50 Bedford Square, London, WC1B 3DP, UK
29 Earlsfort Terrace, Dublin 2, Ireland

BLOOMSBURY, REEDS, and the Reeds logo are trademarks of
Bloomsbury Publishing Plc

First published in Great Britain 2022
Copyright © Oliver Perkins, 2022
Photographs © Oliver Perkins, except where credited on p128
Illustrations pp14, 65, 79, 95 © Dave Saunders
Illustrations pp9, 87, 89, 90 © Tracy Saunders

For legal purposes the credits on p128 constitute an extension
of this copyright page

Oliver Perkins has asserted his right under the Copyright, Designs and Patents
Act, 1988, to be identified as Author of this work

A catalogue record for this book is available from the British Library

ISBN: PB: 978-1-4729-8207-0; ePDF: 978-1-4729-8208-7;
eBook: 978-1-4729-8212-4

2 4 6 8 10 9 7 5 3 1

Typeset in 9 on 11pt Myriad Pro Light by Susan McIntyre
Printed and bound in India by Replika Press Pvt Ltd

To find out more about our authors and books visit www.bloomsbury.com
and sign up for our newsletters

This book is dedicated to Jesus Christ

The heavens declare the glory of God;
The skies proclaim the work of his hands

Psalm 19:1

Thank you very much to my parents, Chris and
Hilary, and my brother Ben, for always supporting
me and for taking many of the photos.

CONTENTS

Introduction to the weather 6

How to use this book 6
A brief introduction to the clouds 6
Deciphering cloud names 7
A bit of history 10

Weather systems 12

Air masses 12
Depressions 13
Anticyclones 14

High clouds 16

Cirrus 16
Thick cirrus (cirrus spissatus) 18
Cirrostratus 20
Cirrocumulus 22
Contrails (cirrus homogenitus) 24

Middle clouds 26

Altocumulus floccus 26
Altocumulus castellanus 28
Altocumulus sheets
    (altocumulus stratiformis) 30
Altocumulus lenticularis 32
Thin altostratus
    (altostratus translucidus) 34
Thick altostratus
    (altostratus opacus) 36

Lower clouds 38

Cumulus 38
Cumulus fractus 40
Fair weather cumulus
    (cumulus humilis) 42
Towering cumulus
    (cumulus congestus) 44
Cumulus cloud streets
    (cumulus radiatus) 46
Featureless stratus
    (stratus nebulosus) 48
Broken stratus (stratus fractus) 50
Stratocumulus 52
Stratocumulus stratiformis 54
Stratocumulus castellanus 56
Nimbostratus 58
Cumulonimbus calvus 60
Cumulonimbus capillatus 62
Thunderstorms 64

Supplementary features and accessory clouds 66

Supplementary features 66
Asperitas 66
Mamma 68
Fall streaks (virga) 70
Anvil (incus) 72
Funnel cloud (tuba) 74
Shelf and roll clouds (arcus) 76
Kelvin-Helmholtz waves
    (fluctus) 78

Accessory clouds	**80**
Cap clouds (pileus)	80
Pannus	82
Velum	84

Fog **86**

Sea fog (advection fog)	86
Land fog (radiation fog)	88
Upslope fog	90
Frontal fog	92
Sea smoke	94

Clouds outside the troposphere **96**

Nacreous (mother of pearl)	96
Noctilucent	98

Optical phenomena **100**

Rainbow	100
Corona	102
Halo	104
Mock sun / sun dog / parhelion	106
Iridescence	108
Ice rainbow (circumzenithal arc)	110
Sunbeams and crepuscular rays	112
Glory	114

Cloud weather lore **116**

Glossary **122**

Quick cloud identifier **124**

Index **126**

How to use this book

This book is designed to make it as easy as possible for you to identify the clouds and other weather phenomena. Use the cloud identifier at the end of the book, or just flick through the pages looking for the cloud or feature that most closely resembles what you are looking at.

This book will help you not only to name the clouds, but to understand how they were made, when to spot them and what weather they indicate. You will also learn about things that aren't strictly speaking clouds but are spotted in our skies, such as bright colours in clouds and different types of fog.

Each entry has an information box including the height of the cloud above the ground, any associated varieties, species, supplementary features or special clouds. An approximate rarity measure is included for each cloud in temperate regions. However, due to local effects, this rarity measure is extremely subjective.

A brief introduction to the clouds

Before we start actually identifying the clouds, it's important to have a little background knowledge. In this and the next chapter, we'll go through how clouds are made and the type of weather systems found in temperate regions.

Air can only hold a certain amount of water vapour, and once this limit is reached the water condenses, forming clouds. Cold air can hold less water vapour than warm air, so as warm air rises and cools it reaches its condensation point – the point at which clouds are produced. This is the main process through which clouds are produced.

Interestingly, water vapour requires a solid or liquid surface to condense around, so the air can become up to 400 per cent saturated before clouds form, unless there is a condensation nucleus present, which is generally a particle about 1/100th of the size of a water droplet.

Clouds are almost exclusively found in the troposphere, which is the layer of atmosphere closest to the ground. It varies between 6km (3.7 miles) thick in polar regions and 18km (11 miles) thick over the Equator. The troposphere ends at a point known as the tropopause. The tropopause is a temperature inversion, which is where the air starts to get warmer as it gets higher. This acts as a lid to the atmosphere as the relatively cooler air can't rise through it, so cloud growth is stopped.

Deciphering cloud names

The World Meteorological Organization (WMO) classifies clouds into ten main groups, called genera. They are divided into three height levels – high, medium and low.

Cloud level	Cloud type
High clouds (base above 6km/3.7 miles) Made of ice crystals	Cirrus (Ci) Cirrocumulus (Cc) Cirrostratus (Cs)
Medium clouds (base between 2 km/ 1.2 miles and 6km/3.7 miles). Made of ice crystals, water droplets or a mix of both	Altocumulus (Ac) Altostratus (As) Nimbostratus (Ns)
Low clouds (base below 2km/1.2 miles) Generally made of water droplets, but can be ice crystals or a mixture of both in cold regions	Stratocumulus (Sc) Stratus (St) Cumulus (Cu) Cumulonimbus (Cb)

The WMO then splits its genera into smaller groups, called species, which describe the shape and internal structure of the cloud. Species can be exclusive to a single genus or common to several genera. There are 15 different species, for example:

◆ Fibratus (meaning straight filaments of cloud)

◆ Humilis (meaning flat cumulus clouds)

Different genera and species can then be split further into nine different varieties, which describe the transparency

Deciphering cloud names

and arrangement of the cloud. Each cloud can have more than one associated variety. Examples of varieties are:

◆ Translucidus (meaning clouds that the sun can be seen through)

◆ Duplicatus (meaning patches of clouds in two or more layers)

There are eleven types of supplementary features, which are additional features attached to the main cloud, such as:

◆ Incus (meaning an anvil top of a cumulonimbus cloud)

◆ Asperitas (chaotic wave-like features at the base of a cloud)

Clouds can also have accessory clouds, which depend on other clouds for their existence. There are four types of accessory clouds, including:

◆ Pannus (fragmented clouds formed by rain)

◆ Pileus (small horizontal 'cap' clouds above cumulus or cumulonimbus clouds)

There are also six types of special clouds, which form due to certain specific local factors. Clouds caused by human factors are followed by the name homogenitus. The special cloud name always follows the main name, for example:

◆ Homogenitus in cirrus homogenitus (aircraft condensation trails)

◆ Flammagenitus in cumulus congestus flammagenitus (clouds formed by convection initiated by heat from fire or volcanic activity)

The main cloud types are all found in the troposphere. However, there are much rarer clouds outside the troposphere, including noctilucent and nacreous clouds. These clouds aren't associated with any weather. We discuss these further on pages 96–99.

Deciphering cloud names

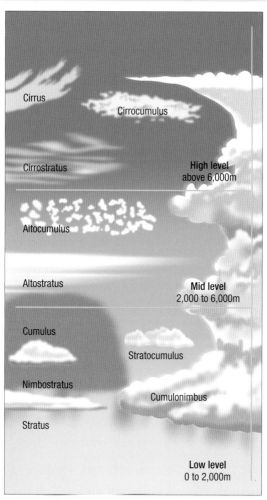

Cirrus

Cirrocumulus

Cirrostratus

High level
above 6,000m

Altocumulus

Altostratus

Mid level
2,000 to 6,000m

Cumulus

Stratocumulus

Nimbostratus

Cumulonimbus

Stratus

Low level
0 to 2,000m

The 10 Cloud Groups.

9

A bit of history

You may be wondering how these clouds were named. Unlike other features of nature, clouds have only been catalogued and named recently.

In 1803, an amateur meteorologist from London named Luke Howard came up with the first cloud classification system. Initially, Howard considered clouds to be part of only three families (cumulus, stratus and cirrus). The scientific community quickly made use of the system, spreading Howard's classifications throughout the world.

Over time, Howard's original system was refined, the first change being that his cumulo-stratus cloud became a stratocumulus cloud, as it was considered to belong primarily in the layered strato group, rather than the heaped cumulo group. The next change was made by Émilien Renou, a French meteorologist. He suggested that clouds are sorted into three height levels, and he introduced two new clouds (altostratus and altocumulus).

Towards the end of the 19th century, Howard's careful cloud classification system was becoming confused by overseas observatories introducing different name variations

---

Where clouds get their names from:

Most cloud names are a combination of these Latin terms:

Cumulus = heap or pile

Stratus = layer

Cirrus = curl of hair

Alto = high (but confusingly refers to medium-height clouds)

Nimbus = rain

Cloud study by Luke Howard c.1808.

for different clouds. This reached a point where the same term was being used for completely different cloud structures or heights. To solve this, a cloud committee was created in 1891, which adopted a ten-cloud classification system. The International Year of Clouds took place in 1896, when the first International Cloud Atlas was created, documenting all the cloud types. Cloud classification improved throughout the 20th century, with variations of different cloud types added.

Did you know?

The International Cloud Atlas was updated as recently as 2017, when one new cloud species was added (volutus), as well as five new supplementary features, five new special clouds and one new accessory cloud.

Air masses

Much of the world's weather, especially in temperate regions, is caused by enormous areas of lower or higher air pressure. These large-scale weather systems are known as depressions and anticyclones. They are caused by different air masses and bring all kinds of clouds, which we will explore later.

Air masses

Air masses are an important concept to understand: they are bodies of air with constant humidity and temperature. These air masses are hundreds or even thousands of miles across. There are usually short transition zones known as a front between each air mass. Air masses get their characteristics by staying over an area for weeks before moving away from their source location. These air masses often meet each

Types of air mass

Air masses are split into two categories – continental (formed over the land) and maritime (formed over the sea).

There are also three different air mass locations, depending on temperature:

Arctic: formed over the arctic regions, such as Greenland and Antarctica.

Polar: air masses formed a bit further from the poles, such as Canada or the North Pacific Ocean.

Tropical air masses: formed over the tropics.

Some meteorologists would also add another category called equatorial air masses.

These air mass locations are each split into the continental or maritime categories – for example, maritime polar or continental tropical.

other in the temperate regions and cause the changeable weather that places like Britain are known for.

Maritime air masses often bring cloudy or rainy conditions. This is because they are much more moist than continental air masses, which often bring dry conditions with higher temperature extremes. In the UK, extreme heatwaves and the freezing 'Beast from the East' are caused by continental air masses.

Depressions

Depressions, as their name suggests, bring poor weather, such as rain, extensive cloud and strong winds. Air rises in a depression, leaving behind less air than the surrounding area, resulting in lower air pressure. They form in the Atlantic when a warm moist maritime tropical air mass from the south meets a cooler and drier maritime polar air mass from the north. As cooler air is denser, it slides underneath the warmer air, rather than mixing. The point at which the air masses meet is known as a front.

Fronts always slope at a shallow angle. In a warm front, this means that the warm air is often over the cold air for a

Types of front

Warm front: Warm air moves towards cooler denser air and rises over it at an extremely gentle gradient. This brings steadily increasing rain for a few hours, which can become heavy.

Cold front: Cold air undercuts warmer air, causing strong, short-lived downpours.

Occluded front: The cold front catches up with the warm front and completely lifts the warm air in the middle. This can bring drizzle, showers and light rain.

Anticyclones

distance over 1,000km (600 miles) before the front actually arrives on the ground. This causes a distinct pattern of clouds that can be recognised as a depression approaches, as seen in the diagram below. A cold front often follows the warm front, and it is about twice as steep.

Anticyclones

Anticyclones are the opposite of depressions: they are areas of higher pressure due to descending air. This descending air suppresses cloud formation, often leading to clear skies, light winds and settled weather. In summer, anticyclones bring hot and dry weather, whereas in the winter they bring cold and dry weather, as there aren't many clouds to trap in the heat from the day.

Anticyclones often cause temperature inversions, where the higher air is warmer than the air below it. These conditions are very stable as the air doesn't mix as it would in a depression. This can either leave days of clear skies or days of stratus and stratocumulus clouds. These stable conditions can trap dust and pollutants, resulting in poor visibility and cloudy days, known as 'anticyclonic gloom'.

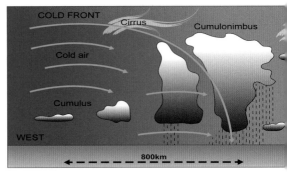

Cross-section of a depression.

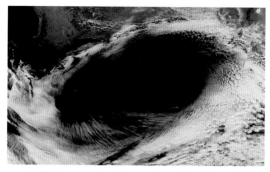

ABOVE: *Anticyclone over Australia.* BELOW: *Anticyclonic gloom.*

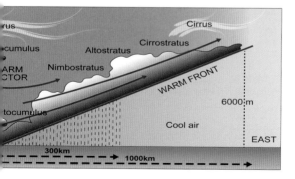

High clouds

High clouds all have their base above about 6km (3.7 miles). The boundary is lower in polar regions and higher in equatorial regions. They are made from tiny ice crystals suspended in the atmosphere. They also never produce any precipitation that reaches the ground. There are three genera of high clouds: cirrus, cirrostratus and cirrocumulus.

Cirrus

Height:	5,000–14,000m (15,000–46,000ft)
Species:	Fibratus, uncinus, spissatus, floccus, castellanus
Varieties:	Duplicatus, intortus, radiatus, vertebratus
Supplementary features:	Fluctus, mamma
Rarity:	2/5

Cirrus clouds are, like all the high clouds, made of ice crystals. They are usually either straight lines (fibratus), have sharp hooks (uncinus), or can be a dense tangled mass (spissatus). Other species are much rarer.

When to spot them

Cirrus clouds are often formed when relatively dry air rises, usually due to an approaching warm front. This causes the water vapour in the air to sublimate (where the vapour 'condenses' straight into ice crystals, without ever being a liquid).

Associated weather

Just to be confusing, these can either mean lovely weather or a storm! However, if the cirrus starts to spread out and thicken, you can be fairly sure that a warm front is on its way, bringing poor weather. If they stay fairly similar or reduce in number then the fine weather should continue.

Cirrus uncinus (below) have a distinct hook on the upwind side, with long straight streamers on the downwind end. The sharper the hook, the quicker the wind direction changes aloft. If the streamers are horizontal then there are very strong high winds, probably due to the jet stream. The stronger these hooks, the more likely it is that a depression is coming and the more vigorous it will be.

Thick cirrus (cirrus spissatus)

Height:	5,000–14,000m (15,000–46,000ft)
Varieties:	Duplicatus
Rarity:	2/5

The Latin word spissatus means 'thickened', and only cirrus clouds have this feature. They are so thick that they block sunlight and can appear dark grey from below, so don't confuse them with an altostratus cloud. They frequently exhibit optical phenomena, such as mock suns on the edges.

When to spot them

They are often seen after cumulonimbus clouds. In this case, they are the leftover anvil of a decayed cumulonimbus cloud and are known by the name cirrus spissatus cumulo-nimbogenitus (opposite top). They are also found when cirrus clouds thicken and merge together. This is particularly common in the fast high-altitude winds known as the jet streams. In this case cirrus clouds form into long streaks or banners that converge at the horizon (opposite bottom).

Associated weather

They can be seen before warm fronts. However, to confirm this, make sure the clouds are thickening first and that the cloud base is lowering as they often form at other times. If the clouds are thickening before a warm front, rain is expected in 8–36 hours. If the long streaks of jet stream cirrus spissatus are seen then expect a particularly vigorous depression with gale-force winds.

Cirrostratus

Height:	6,000–13,000m (20,000–45,000ft)
Species:	Fibratus, nebulosus
Varieties:	Duplicatus, undulatus
Rarity:	2/5

Cirrostratus clouds are transparent sheets of cirrus clouds that often go unnoticed. The sun or moon is always visible through cirrostratus. They exhibit many optical phenomena, most commonly haloes, but also mock suns and circumzenithal arcs.

There are two species of cirrostratus clouds. One is cirrostratus fibratus, fibres of cirrostratus clouds that cover a large portion of the sky (opposite top). These are often individual cloud strands that look like wisps of hair. They are much lighter and more transparent than ordinary cirrus clouds. The other species, cirrostratus nebulosus (opposite bottom), is a uniform veil of cirrostratus that usually covers the whole sky. It is completely featureless and is often only noticeable due to the faint halo around the sun or moon.

When to spot them
They are often the first clouds spotted before a warm front and will slowly thicken and lower into an altostratus cloud as the front approaches.

Associated weather
As they often form before a warm or occluded front, they can mean rain or snow in 10–20 hours. As they thicken, the sun starts to lose some of its heat, which is when we would ordinarily start to notice cirrostratus cloud.

Cirrocumulus

Height:	5,000–13,000m (15,000–45,000ft)
Species/varieties:	Stratiformis, lenticularis, floccus, castellanus
Varieties:	Lacunosus, undulatus
Supplementary features:	Cavum, mamma, virga
Rarity:	3/5

Cirrocumulus clouds are patches or large groups of tiny, regularly spaced cloudlets. They are also known as 'mackerel skies' as they look like fish scales. They are made up of ice crystals or supercooled water (where the water is cooled well below freezing but is still a liquid).

They are easy to confuse with altocumulus clouds, but the cirrostratus clouds have no shading and are either light blue or white. The sun can always be seen through them. Another way to work out which one you're seeing is simply to raise your fingers to 45 degrees at arm's length and measure the width of the cloudlets on your fingers. If the cloudlets are only one finger wide, then they are likely to be cirrocumulus, while altocumulus clouds generally cover two to three finger widths. Interestingly, cirrocumulus cloudlets can be up to 200m (650ft) across, which is the same size as the larger-looking altocumulus cloudlets, but because of their height they look smaller.

When to spot them
They can be seen 10–15 hours before a warm front approaches, as they require moist air to form.

Associated weather
If seen before a warm front, rain and stronger winds are to be expected in the few hours before a front arrives.

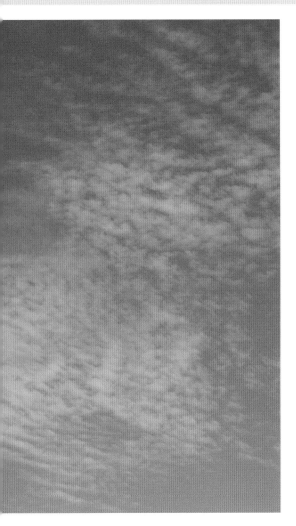

Contrails (cirrus homogenitus)

Height:	7,500–13,000m (25,000–45,000ft)
Varieties:	Vertebratus
Rarity:	1/5

Condensation trails (contrails) are long lines of cloud formed behind an aircraft that persist for at least ten minutes but can last for hours. They are formed by the condensation of water vapour around aeroplane exhaust particles. The air must be well below freezing and be moist enough for the contrails to appear, but not so moist that there are already other clouds in the sky. The water vapour in the air cannot condense into clouds without particles known as nuclei, which act as centres for the air to condense around. Contrails are often mentioned in the media as they trap more heat in the atmosphere than the sun they reflect, contributing to global warming.

When to spot them

Contrails are spotted on clear days throughout the year, especially in major flight paths. They are seen either during anticyclones (times of high pressure and otherwise clear skies) or before a warm or occluded front. When there are strong upper winds they can transfer into more natural cirriform clouds. Here the special cloud name homomutatus is added, which means caused by human activity.

Associated weather

Contrails don't form any weather at ground level; however, if they are especially thick and long-lasting, they can indicate the passage of a warm or occluded front in 12–48 hours.

Contrails

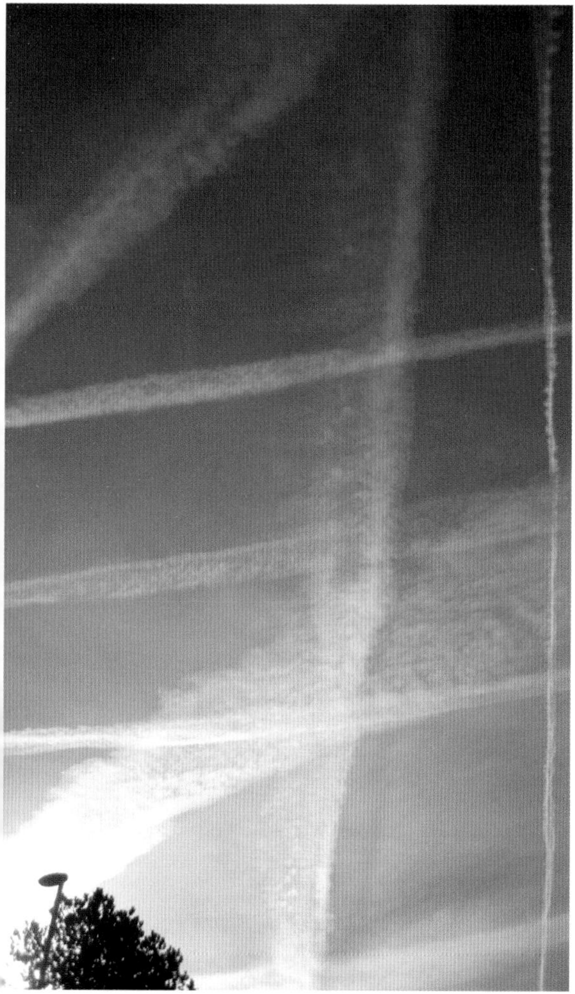

Middle clouds

The bases of the middle clouds are usually between 2km (1.2 miles) and 6km (3.7 miles) high. The middle clouds are made up of ice crystals or water droplets, or often a mixture of the two. The three middle cloud genera are altocumulus, altostratus and nimbostratus (which looks like lower cloud from the ground, so is included in that section).

Altocumulus floccus

Height:	2,000–6,000m (6,500–20,000ft)
Varieties:	Duplicatus, radiatus
Supplementary features:	Virga
Rarity:	3/5

Altocumulus floccus clouds are small tufts of cloudlets that are said to resemble a flock of sheep. They are light grey or white lumps and are often found in scattered disorganised patches. They often exhibit fibrous trails of ice crystal virga. Their cloudlets are smaller than would usually be expected with altocumulus. They have ragged bases and often a small bulge at the top.

When to spot them
These are spotted before storms, alongside altocumulus castellanus (see next page), or even sometimes after

This innocent-looking cloud was spotted just a few hours before over 250,000 lightning strikes were recorded in the English Channel.

the base of altocumulus castellanus clouds dissipate. All altocumulus clouds seem from the ground to be smaller than the elements of stratocumulus clouds, which always measure more than 5 degrees across (three fingers at arm's length).

Associated weather

Altocumulus floccus clouds, like their cousins altocumulus castellanus, are found when the air is unstable (when air is warmer than the air above it). This means that thunderstorms are likely as it allows small cumulus clouds to increase significantly in size once they reach the unstable layer, often allowing them to become full cumulonimbus storm cells.

Altocumulus castellanus

Height:	2,000–6,000m (6,500–20,000ft)
Varieties:	Duplicatus, radiatus
Supplementary features:	Virga
Rarity:	3/5

Altocumulus castellanus are named after the 'towers' that rise above the base of the cloud. These turrets are often found in lines or bands. Like cumulus clouds, altocumulus castellanus are caused by air heated from below, which rises into the cooler air above. This means that the atmosphere is unstable. What separates these from normal cumulus clouds is that the instability in the atmosphere starts in the middle layer of the troposphere. These may produce precipitation, but it will almost always evaporate before it hits the ground, so they don't cause poor weather themselves.

When to spot them

Since these clouds show instability in the middle layers they are a particular favourite of storm chasers. This is because they often precede huge convective clouds such as cumulonimbus, which cause heavy showers or thunderstorms. The instability causes small cumulus clouds to grow into enormous cumulonimbus clouds.

Associated weather

The larger the towers, the more unstable the atmosphere, and the gustier and larger the thunderstorms are likely to be. It can take up to 24 hours for thunderstorms to appear after these clouds, but they are usually a fairly reliable indication of some exciting weather!

Altocumulus sheets (altocumulus stratiformis)

Height:	2,000–7,000m (6,500–21,000ft)
Varieties:	Duplicatus, lacunosus, opacus, translucidus, perlucidus, radiatus, undulatus
Supplementary features:	Asperitas, cavum, fluctus, mamma, virga
Rarity:	2/5

Altocumulus in extensive flat layers is the most common altocumulus species. They can take many forms, such as long billows, or lots of cloudlets arranged regularly, like bread rolls. The elements can either be completely separate from each other, with blue sky in between, or merged together with few gaps, but with clear thicker and thinner sections. They have flat bottoms and are found in thin layers either covering the whole sky or in a few separated patches. They are often caused by 'waves' of humid air, which is rising and condensing in the clouded areas and creating gaps between the cloudlets where the air sinks.

When to spot them

Altocumulus stratiformis is often seen before a weakening warm front. This is because the altostratus cloud that would usually be seen has fragmented, leaving patches of altocumulus stratiformis perlucidus between the altostratus layers. It is also often present in the warm sector (the section between warm and cold fronts); however, it is difficult to see as it is usually obscured by lower clouds.

Associated weather

It usually doesn't indicate any change in the weather, but if it rapidly thickens and lowers then it is likely to be seen before a weak depression, resulting in patches of rain and extensive cloud.

Altocumulus lenticularis

Height:	2,000–5,000m (6,500–15,000ft)
Varieties:	Duplicatus
Rarity:	4/5

These well-defined lens- or almond-shaped clouds are a spectacular sight and are often mistaken for UFOs. They are caused by a moist airflow being disrupted by mountains or hills. This disruption causes standing waves in airflow, with it rising in some places and sinking in others. The lenticular clouds are formed in these areas of rising air. In some conditions, multiple lenticular clouds are created at the crest of each of these waves of air. Sometimes the clouds are visible even when observed from well out of sight of the hills or mountains that formed them. Although the air itself is moving, the clouds stay stationary and can persist for long periods, assuming the wind strength and direction stay the same.

When to spot them
These are seen downwind of hilly or mountainous areas, where they can be relatively common. However, for people who live well away from these areas they are a very rare and exciting cloud to spot.

Associated weather
There are no specific weather conditions associated with these clouds.

Did you know?

Experienced glider pilots love lenticular clouds as the rising air is often strong and smooth. The world records for gliding height and distance were set using lift from these clouds.

Thin altostratus (altostratus translucidus)

Height:	2,000–7,000m (6,500–20,000ft)
Varieties:	Duplicatus, radiatus, undulatus
Supplementary features:	Mamma
Rarity:	1/5

Altostratus is a fairly featureless white or grey cloud. It is pretty unexciting and causes dull, overcast conditions. These clouds are so boring that they have no species associated with them and very few varieties. This entry is looking specifically at altostratus translucidus, the variety where the sun and even sometimes the moon can be seen through the cloud. However, the sunlight becomes diffused and it looks as though it is being seen through frosted glass.

On occasion, there are parallel streaks visible on the bottom of the altostratus cloud. This is known as altostratus radiatus, but is quite unusual to spot. An altostratus translucidus cloud sometimes shows optical phenomena such as coronae and iridescence when it is made of uniformly sized water droplets. However, often altostratus is made up of a mix of water droplets and ice crystals, which means optical phenomena are extremely unlikely.

When to spot them

Altostratus usually forms before depressions. It often starts off as a cirrostratus cloud, which slowly lowers and thickens as a warm or occluded front comes nearer. The altostratus cloud will initially be very thin, making it altostratus translucidus, where the sun can be seen through it. However, as the front draws closer the altostratus will thicken to altostratus opacus (see next page).

Associated weather

If cirrus clouds are spotted before the arrival of the altostratus cloud, or if the altostratus steadily thickens to altostratus opacus then it is likely a depression is coming. If so, the wind is likely to pick up significantly and rain will be only a few hours away as the altostratus thickens into nimbostratus.

Thick altostratus (altostratus opacus)

Height:	2,000–7,000m (6,500–21,000ft)
Varieties:	Duplicatus, radiatus, undulatus
Supplementary features:	Praecipitatio, virga, pannus, mamma
Rarity:	1/5

Even more dull than altostratus translucidus is altostratus opacus. This cloud can be thousands of metres thick, and it completely blocks out the sun or moon so that its outline can't be seen and no shadow is cast on the ground. Altostratus opacus has no optical effects associated with it as it is too thick for the sun's rays to penetrate without being reflected in every direction off the water droplets or ice crystals.

When to spot them

Occasionally, the upper parts of a cumulonimbus cloud or a thunderstorm system may spread out, producing an altostratus cloud (usually of the altostratus opacus variety). However, altostratus opacus is usually spotted before a depression, where altostratus translucidus (see previous page) thickens into altostratus opacus.

Associated weather

Before a depression altostratus will thicken, eventually becoming a rain-bearing nimbostratus cloud. In a depression the rain will become heavier until the warm or occluded front passes through. The wind will also increase steadily.

Lower clouds

Lower clouds usually have their base below 2km (1.2 miles). They are almost always made of water droplets, except on extremely cold days. There are four genera: cumulus, stratus, stratocumulus and cumulonimbus.

Cumulus

Height:	200–2,000m (650–6,500ft)
Species:	Fractus, humilis, mediocris, congestus
Varieties:	Radiatus
Supplementary features:	Fluctus
Rarity:	1/5

Cumulus clouds are extremely common and are the classic 'cotton wool' cloud. They are caused by convection, where air rises in updraughts due to the heating of the sun. As the air rises, it cools and then reaches a point (the condensation point) where it can no longer hold the water as vapour, so it condenses into cloud. There is a relatively even level in the sky where the water vapour first condenses, so the clouds almost always have flat bottoms at the same level as all the other clouds around them. They have the bulging cotton-wool-shaped tops as the updraught is strongest in the middle, so the moist air rises higher.

Cumulus clouds come in all different shades of white and grey, but this is mainly due to the position of the sun, rather than how likely they are to rain. If the sun is behind the clouds it will shade the side nearest you, resulting in it looking darker than when the sun is shining straight on the cloud, where it would look a brilliant white (see the undersides of the cloud in the photo opposite when compared to the tops).

There are different species of cumulus clouds depending on their height, and these are covered in the next pages.

Cumulus fractus

Height:	200–2,000m (650–6,500ft)
Varieties:	Radiatus
Supplementary features:	Fluctus
Rarity:	1/5

Cumulus fractus clouds are small cumulus clouds with very ragged edges, which are constantly and quickly changing in form. They have irregular patterns, like torn pieces of candyfloss. They don't usually have flat, clearly defined bases, but often have a slightly bulging top.

When to spot them

The Latin word fractus means 'fraction' or 'small part', so cumulus fractus are usually the first parts or wisps of cumulus clouds that are just starting to develop into cumulus humilis clouds in the early morning on an otherwise clear day. The other time to spot them is in the evening as the sun starts to set and the cumulus clouds start to dissipate, often looking spectacular in the evening light. Occasionally, cumulus clouds will be formed by a particular heat source that's causing them to grow. In this case, the cumulus clouds may also decay into cumulus fractus clouds when they move away from the heat source.

Associated weather

If seen in the morning, it could mean anything from a lovely day, with a few cumulus humilis clouds, to full-blown thunderstorms in the afternoon. You'll have to keep an eye on whether the cumulus fractus clouds develop into tall cumulus congestus clouds, or stay as flat cumulus humilis clouds.

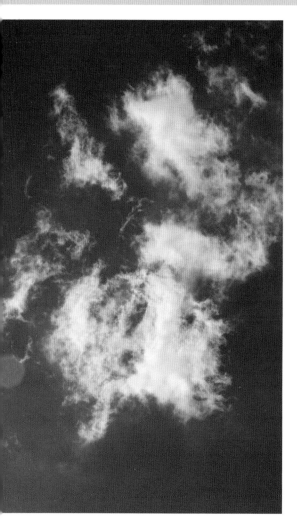

Fair weather cumulus (cumulus humilis)

Height:	200–2,000m (650–6,500ft)
Species/varieties:	Radiatus
Supplementary features:	Fluctus
Rarity:	1/5

Cumulus humilis clouds are cumulus clouds that are much wider than they are tall. They are known as fair weather cumulus clouds as their presence after mid-morning indicates fair weather for the rest of the day.

When to spot them

Cumulus humilis clouds have a daily cycle, which is very common, especially in the summer months in temperate regions. Early in the morning, the sky is completely clear, but as the morning progresses small cumulus fractus clouds form. These gradually merge into small cumulus humilis clouds, which grow to their largest size in the early afternoon, often becoming cumulus mediocris or cumulus congestus clouds. As the strength of the sun decreases, the updraughts supporting the clouds decrease, so they start to reduce in size and become the more fragmented cumulus fractus clouds. Just after sunset, the sky is usually completely clear again.

On some days, the growth of cumulus clouds will be restricted by a layer of warmer air, known as a temperature inversion. In this case, the clouds can only grow outwards rather than up, and sometimes merge into other cumulus humilis clouds. This can cause them to develop into a layer of stratocumulus cloud (where there are much smaller gaps between individual clouds).

Associated weather

If the convection is particularly vigorous, the cumulus humilis clouds can grow vertically into cumulus mediocris (where the height is approximately equal to the width) or even cumulus congestus clouds, which could then bring showers. However, usually they only bring fair weather.

Towering cumulus (cumulus congestus)

Height:	200–6,000m (650–20,000ft)
Varieties:	Radiatus
Supplementary features:	Tuba, virga, praecipitatio
Accessory clouds:	Pannus, pileus, velum
Rarity:	2/5

Cumulus clouds that are taller than they are wide are known as cumulus congestus clouds. They have bright white cauliflower-like florets on the top of them, unlike cumulonimbus clouds, which will have smooth or fibrous cirrus-like edges as the water in the clouds has frozen due to the high altitude.

When to spot them

In temperate regions, these rarely form outside of the summer months; however, in tropical regions they are seen all year round. If the sky is clear on a summer morning, but some cumulus clouds start to develop quickly as the sun strengthens, then there may well be cumulus congestus clouds – you'll just have to keep an eye out to see if the cumulus clouds start growing vertically.

Associated weather

If you are in hilly or mountainous regions you will find that these clouds grow vertically, potentially bringing rain as they go uphill, but as they descend downhill they'll decay and are much less likely to bring rain, even if they were raining further up the hill.

In tropical regions they will be even taller, so are more likely to rain. In temperate regions cumulus congestus often won't rain itself, but may herald the development of rain-bearing cumulonimbus clouds later in the day.

Cumulus cloud streets (cumulus radiatus)

Height of base:	200–2,000m (650–6,500ft)
Species:	Fractus, humilis, mediocris, congestus
Rarity:	2/5

Cumulus clouds are often arranged in regularly spaced lines parallel to the wind direction, known as cloud streets. They can extend for hundreds of miles. They seem to merge on the horizon due to perspective.

When to spot them

They are caused by uneven solar heating of the ground, where some areas are warmed much more significantly than others, causing strong thermals of rising air. The wind carries the cloud downwind, allowing another cloud to be formed by the same thermal. Occasionally there is just one cloud street, which would have been caused by

a particularly strong source of thermals, such as a hillside, airport runway or often a single isolated island.

They are made of rolls of convection, with rising air forming clouds and falling air suppressing cloud formation in the gaps. These convective rolls fit together like cogs and are generally spaced a few kilometres apart.

Associated weather

Cloud streets are very useful for gliders as they can make use of the updraughts to stay in the air. Gliders can follow cloud streets for hours, hence their name.

Sailors use cloud streets too, as there is stronger wind in between the clouds than under them. This is because the falling air between the clouds has not been slowed by friction with the Earth's surface as much as the rising air underneath the clouds.

Featureless stratus (stratus nebulosus)

Height:	0–2,000m (0–6,500ft)
Varieties:	Opacus, translucidus
Supplementary features:	Fluctus, praecipitatio
Rarity:	1/5

Stratus nebulosus is a low featureless grey cloud that often covers huge areas. They are usually no more than 500m (1,500ft) high, so they often hide the tops of buildings or hills (as seen in the photo). Its base will sometimes descend down to ground level, in which case there is no fundamental difference between fog and stratus clouds (although technically fog is not considered a cloud). Stratus clouds don't experience any optical phenomena.

When to spot them

During the winter, 'anticyclonic gloom' is often experienced, where vast areas of stable air trap stratus clouds, often for days on end. They can also build north of depressions (or south in the southern hemisphere).

Associated weather

They can bring light drizzle, but often just occur on dry but humid days. They usually last for quite a few hours or even days.

Broken stratus (stratus fractus)

Height:	0–2,000m (0–6,500ft)
Supplementary features:	Fluctus
Rarity:	2/5

Stratus fractus clouds are ragged shreds of broken stratus clouds. They are very low clouds that move quickly and change shape continuously. They can either form as separate shreds or be below a precipitating cloud such as altostratus or nimbostratus, in which case they are often known as pannus clouds (see page 82). Stratus fractus clouds can grow in number and size, and eventually merge, covering the whole sky, or they can stay as a few ragged shreds.

When to spot them

It can be difficult to tell the difference between stratus fractus and cumulus fractus. Stratus fractus is usually darker and has a smaller vertical extent, but without having the clouds side by side it is hard to work this out. The best way of telling the difference is that cumulus fractus clouds are formed by convection, and form as a cumulus cloud starts developing and when it decays, while stratus fractus is caused when warm air passes over a colder area of sea or land.

Associated weather

Stratus fractus don't produce rain themselves, but they will appear to be raining when below a raining nimbostratus or altostratus cloud.

Stratocumulus

Height:	400–2,000m (1,300–6,500ft)
Species:	Stratiformis, cumulogenitus, castellanus, lenticularis
Rarity:	1/5

Stratocumulus clouds are the most common cloud type, especially over the oceans. They have features from both stratus and cumulus clouds. They are usually arranged in one layer, with small gaps between the elements. The cloud elements themselves are always larger than 5 degrees across (when measured from 30 degrees above the horizon). They can display all kinds of features, such as rolls, heaps or flat cloudlets. They can cover hundreds of miles and usually cover the whole sky.

There are two ways that stratocumulus clouds form. The first is from cumulus clouds that can't grow any higher as they hit a temperature inversion. Instead, they spread out horizontally, forming stratocumulus. The other way is where a small amount of convection occurs within a stratus cloud, as the tops of stratus clouds cool. This convection causes decay in parts of the cloud, breaking it up into the elements of stratocumulus clouds.

When to spot them
In the winter in temperate regions, temperature inversions within anticyclones may form huge areas of stratocumulus cloud. They aren't typically seen in depressions.

Associated weather
They usually indicate that the weather won't change significantly in the near future. They don't bring precipitation and winds around them are usually light.

Stratocumulus stratiformis

Height:	400–2,000m (1,300–6,500ft)
Varieties:	Duplicatus, opacus, translucidus, perlucidus, lacunosus
Supplementary features:	Asperitas, cavum, fluctus, mamma, praecipitatio
Rarity:	1/5

Stratocumulus stratiformis clouds often cover the entire sky. They are extremely varied clouds with large cloud elements and small gaps between the elements. They often have very exciting shapes and can be different every day, such as in comparing this photo to the previous page.

When to spot them

Stratocumulus stratiformis clouds can cover hundreds of miles and are very common, especially over the ocean.

Associated weather

Along with other stratocumulus clouds, it can be formed under a temperature inversion, where the air is trapped from rising by a layer of warmer air. In these situations, the cloud can be very dangerous for air quality as particles from industrial processes cannot escape into the atmosphere. The worst smogs are caused in conditions like these and claim thousands of lives each year through respiratory illnesses.

The sun's outline can sometimes be seen through stratocumulus stratiformis; however, more often than not it is completely obscured, except when it can be seen between the gaps. A common optical feature seen in stratocumulus clouds, including the stratiformis species, is crepuscular rays, which are seen when the sun shines through the gaps between the cloud elements.

Stratocumulus castellanus

Height:	400–2,000m (1,300–6,500ft)
Varieties:	Duplicatus
Supplementary features:	Virga, fluctus
Rarity:	4/5

Very rarely, stratocumulus clouds form under conditions with sufficient convection to grow vertically. These are known as stratocumulus castellanus, due to the turrets that rise above the common base of the stratocumulus cloud. These turrets are sometimes taller than they are wide and give the cloud an appearance of battlements when viewed from the side as they are usually arranged in lines. They are low clouds, are normally dark grey and look as though they should be producing rain.

When to spot them

They are easily confused with ordinary stratocumulus clouds. This is because the flat base of the stratocumulus castellanus cloud usually hides the rising turrets, so the best time to spot them is when viewed side-on, or from an aeroplane. This makes it a very rare cloud to spot, unless you happen to be a pilot!

Associated weather

Stratocumulus castellanus don't produce any rain themselves. However, if they continue to grow vertically they could grow into cumulus congestus stratocumulogenitus, bringing showers, or cumulonimbus stratocumulogenitus, which could bring showers, thunderstorms and hail.

Nimbostratus

Height:	500–6,000m (1,500–20,000ft)
Supplementary features:	Virga, praecipitatio
Accessory clouds:	Pannus
Rarity:	1/5

Nimbostratus clouds are undoubtedly the least liked clouds, unless you happen to be a farmer desperate for rain. They bring hours of constant rain and look extraordinarily dull, usually just a single unenticing shade of grey. Often, the base of the cloud seems to blend in with the landscape around it. Technically, nimbostratus clouds are considered mid-level clouds; however, they can cover all three levels and look like low clouds from the ground, so are included in this section.

When to spot them
In temperate regions, most of the winter rain is from nimbostratus clouds. They are usually formed by a thickening of altostratus cloud during a depression, where they can cover thousands of square kilometres. However, they can also be caused by cumulonimbus clouds spreading out, bringing rain after the main storm system has passed. In this case, they cover much smaller areas.

Pannus clouds are a good indicator of whether the cloud is raining, and thus a nimbostratus cloud, or just a dark stratus or altostratus cloud. If rain is coming from the cloud then it often forms pannus clouds below the main cloud base, which can even thicken until they completely obscure the nimbostratus cloud.

Associated weather
Nimbostratus indicates a prolonged period of rain or snow.

Did you know?

Nimbostratus clouds are the only cloud genera that have no species or varieties.

Cumulonimbus calvus

Height:	500–16,000m (1,500–52,500ft)
Supplementary features:	Mamma, praecipitatio, tuba, virga
Accessory clouds:	Pannus, pileus, velum
Rarity:	2/5

As cumulus congestus clouds grow further, they develop into cumulonimbus calvus clouds, which cover all three cloud layers but are included in this section as their base is low. The tops of the cloud are a brilliant white, whereas light is blocked from reaching the lower layers, which can be a dark grey or even black shade. Each of the round puffy cloud tops is caused by a different thermal, usually powered by heat from the sun.

When to spot them
Cumulus calvus clouds form when cumulus congestus clouds begin to lose the well-defined 'cauliflower' appearance at the top and sproutings on the upper parts become softer and flattened. At this point, the tops of the clouds have frozen into tiny ice crystals. This exact difference can be difficult to see without binoculars.

Associated weather
Cumulus calvus clouds produce heavy rain showers and strong winds that often reach 30–40 knots. The temperature also drops due to the rain dragging the cold air downwards. They can also produce hail and lightning, but this is much rarer. It is usually cumulonimbus capillatus clouds that produce the most ferocious weather with hail and lightning. Cumulonimbus calvus clouds don't have the fibrous or striated parts that cumulonimbus capillatus exhibit and they haven't yet reached the tropopause.

Cumulonimbus capillatus

Height:	500–20,000m (1,500–65,000ft)
Supplementary features:	Arcus, cauda, incus, mamma, murus, praecipitatio, tuba, virga
Accessory clouds:	Fumen, pannus, pileus, velum
Rarity:	2/5

As cumulonimbus calvus clouds develop further, the glaciated (frozen) upper section becomes fibrous. At this point, the cloud has become a cumulonimbus capillatus cloud. The Latin word capillatus, meaning 'hairy', describes the striated or clearly fibrous structure. These are the tallest clouds, reaching up to 20km (65,000ft) from top to bottom in the tropics. This height brings the most ferocious weather, including heavy showers, hail and lightning.

The top often takes the shape of a plume of cirrus, a vast disorderly mass of hair or an anvil. This anvil is caused when the cloud cannot grow any higher as it hits temperature inversion at the top of the troposphere. The anvil is known as cumulonimbus capillatus incus.

When to spot them

The base is usually between 500m and 1,000m (1,500–3,000ft) high and obscures the cloud itself. This can make it difficult to identify from below whether the cloud is cumulonimbus or nimbostratus as they can have similar features such as pannus and are both rain-bearing. It's also impossible to tell the difference between the calvus or capillatus species from below, as it is the top that shows the difference between the clouds.

Associated weather
Cumulonimbus showers are generally strong and come in heavy bursts, unlike the widespread persistent rain from nimbostratus, and if hail or lightning is present, you can be sure it's a cumulonimbus cloud

Thunderstorms

There are four different types of thunderstorms: single-cell, multicell (both in clusters and squall lines) and supercell. An individual cell is a pair of an updraught and a downdraught. We have already looked at single-cell storms, which are individual cumulonimbus clouds. However, they become even more ferocious when more than one cumulonimbus calvus and/or capillatus clouds merge. This is usually due to stronger winds higher up (wind shear), which allow new cells to join the thunderstorm, unlike vertical single-cell storms, which have a downdraught that suppresses new cells from growing.

Multicell thunderstorms are made up of multiple cells at different stages of development, as seen in the diagram below. Each cell takes it in turns to be the most dominant. New cells are formed on the upwind side of the storm, where they start as cumulus clouds, developing into cumulus congestus clouds. In the middle of the storm are the mature cumulonimbus cells, which provide the heavy

Did you know?

It is commonly thought that for every second between seeing lightning and hearing thunder, the storm is a kilometre away. In fact, every three seconds between the lightning and thunder is one kilometre, and five seconds is a mile.

Developing
Cumulus
Congestus

Cross-section of a multicell thunderstorm.

rain, lightning and hail. Each individual cell only lasts for between 20 minutes and an hour, after which they start to dissipate, leaving just the anvil behind, which is called cirrus spissatus cumulonimbus. These multicell clusters produce large hail, flash floods and occasionally weak tornadoes in places like Tornado Alley in central USA.

Multicell thunderstorms can also form as squall lines. Here, new cells are formed on the front edge, producing strong winds coming directly out of the front of the system. Individual thunderstorm cells within the line produce heavy rain and hail as they pass overhead.

The most dangerous type of storm is a supercell storm. This has a large single rotating updraught and can last for 2–4 hours. They can produce large hail and strong winds, and bring tornadoes 30 per cent of the time. These are much rarer, especially in places like Europe. They are most common in Tornado Alley in North America.

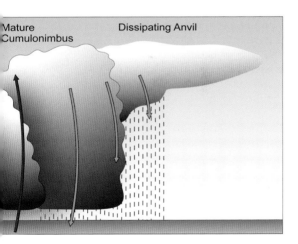

Mature Cumulonimbus

Dissipating Anvil

Supplementary features

There are eleven types of supplementary features, which are additional features attached to the main cloud. There are also four types of accessory clouds, which depend on other clouds for their existence. Accessory clouds can be separate from the main cloud body, or partly merged with it. This section only includes the supplementary features and accessory clouds that you are likely to see. In addition, if you see rain, snow or hail falling from a cloud and reaching the ground or sea, then that technically is a supplementary feature called praecipitatio.

Asperitas

Height:	400–7,000m (1,300–21,000ft)
Parent clouds:	Altocumulus, stratocumulus
Rarity:	5/5

Asperitas clouds look like the underside of waves in the sea. They look spectacular, especially when illuminated from the sun, and are extremely rare. So rare that it was only added to the International Cloud Atlas as a supplementary feature in 2017 after it was proposed by Gavin Pretor-Pinney, founder of the Cloud Appreciation Society. He suggested it be known as asperitas, meaning 'roughness'. The previous cloud to be added to the Atlas was cirrus intortus in 1951, so the introduction of this cloud was big news at the time.

Where to spot them
The cloud is so newly classified that meteorologists aren't entirely sure how it is formed. One theory is that it is formed when

winds shear mammatus clouds into wavelike forms. We do know that asperitas clouds only form in unstable conditions.

Associated weather

Asperitas doesn't form rain itself; however, they are usually spotted after thunderstorms, though rarely before them. They have also been spotted in very calm environments. What is clear is that they look very ominous, almost apocalyptic, but whether or not they bring thunderstorms and other poor weather is still unknown.

Mamma

Height:	Any
Parent clouds:	Cumulonimbus, altostratus, cirrus
Rarity:	4/5

Mamma look like lobes or udders and most often form under the base of cumulonimbus clouds, although they also form under altostratus and cirrus clouds. Mamma comes in all kinds of different shapes and sizes, from spherical pouches or lobes to globules arranged as connected cells. The lobes themselves are usually made from ice and are around 500m deep (1,500ft) and 2,000m wide (6,500ft).

Where to spot them

Mamma are caused by strong downdraughts, where pockets of cooler moist air sink rapidly against the main pattern of upward movement of warm humid air. The exact mechanism as to how this happens is still unknown. Each individual lobe usually only lasts for around ten minutes; however, the cluster of them can last for up to a few hours.

Mamma are very dangerous for pilots due to the huge amount of turbulence associated with the air around them.

Associated weather

If there hasn't already been a thunderstorm or heavy shower then one may be coming imminently; however, these clouds are usually seen after a shower, so aren't very helpful in forecasting weather.

Did you know?

The name mamma is derived from the Latin word for 'breast' or 'udder'.

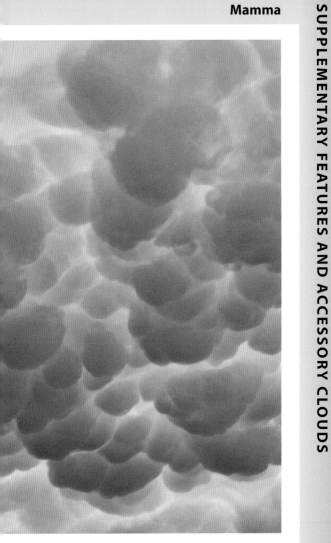

Fall streaks (virga)

Height:	Any
Parent clouds:	Cirrus, cumulus congestus, cumulonimbus, stratocumulus, nimbostratus, altocumulus castellanus, altocumulus floccus
Rarity:	4/5

Virga are trails or streaks of precipitation that evaporates before it hits the ground. Virga, from the Latin for 'branch' or 'rod', are so called due to their often elongated shaft shape. When falling from higher clouds virga are made from ice crystals, the most common example being the fallstreaks from cirrus clouds. The phenomenon is most often seen from lower clouds such as cumulus congestus, cumulonimbus, stratocumulus and nimbostratus clouds, where it is usually rain falling rather than ice. In these cases, it can look particularly spectacular at sunset when illuminated from the side.

The trails are also often seen from altocumulus castellanus and altocumulus floccus, in which case they can be referred to as 'jellyfish clouds', due to the streaks below the top. The reason the rain doesn't hit the ground is often because it heats up and evaporates as the air pressure increases nearer the ground. This phenomenon is called compressional heating.

Where to spot them
They are particularly common in deserts but are also often seen in temperate climates, especially below cirrus clouds.

Associated weather
They can indicate fine weather (in some cirrus clouds), or imminent rain (under nimbostratus or cumulonimbus clouds), so it's best to check which cloud they are associated with to see what weather they will bring.

Don't confuse it with the supplementary feature praecipitatio, which is rain, hail or snow that reaches the ground, unlike virga.

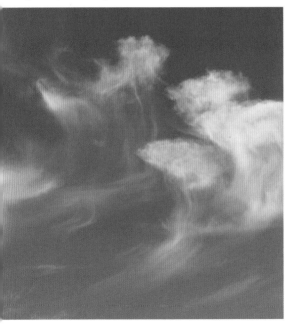

SUPPLEMENTARY FEATURES AND ACCESSORY CLOUDS

Anvil (incus)

Height:	5,000–16,000m (15,000–52,500ft)
Parent clouds:	Cumulonimbus capillatus
Rarity:	3/5

If cumulonimbus capillatus keep growing vertically until they reach the tropopause, their tops will have no choice but to spread out sideways as they can't rise through the temperature inversion. This creates an anvil top known as the incus (Latin for 'anvil'). It can either have a smooth appearance as in the top photo (opposite), or it can be very fibrous or striated.

When to spot them

Incus clouds are always made of ice crystals and are only ever associated with cumulonimbus capillatus clouds. Mamma clouds are often spotted on the underside of the anvil.

Associated weather

An anvil usually indicates that the cumulonimbus cloud is beginning to decay, but will still be raining quite heavily. Sometimes the anvil can separate from the main cloud, which can produce lightning miles from the cumulonimbus cloud itself. Once the cumulonimbus cloud has dissipated, it can leave the anvil behind. This is then considered to be cirrus spissatus cumulonimbogenitus (page 19), which appears very dark in the sky. These can persist as cirrus clouds for days.

Funnel cloud (tuba)

Height:	0–2,000m (0–6,500ft)
Parent clouds:	Cumulonimbus, cumulus congestus
Rarity:	4/5

A tuba forms when a rotating column of air descends below the main cloud base, condensing moisture into water vapour. They are funnel-shaped clouds that form beneath either cumulonimbus or cumulus congestus clouds.

When to spot them

They are more common than you would expect and can even form under calmer cumulus clouds if the conditions are perfect, but this is highly infrequent. A funnel cloud usually lasts for no more than a few minutes, after which the vortex collapses.

Associated weather

Rain, hail and thunder are associated with cumulonimbus clouds that bring these, although it may stay dry if seen below cumulus congestus clouds.

Rarely, a tuba will touch the ground, causing it to be known as a landspout, or a waterspout if it is over a body of water. These are not nearly as damaging as tornadoes, but are still quite dangerous. Tubas can also develop into tornadoes, but these are caused under the very different conditions of a supercell thunderstorm.

Shelf and roll clouds (arcus)

Height:	100–2,000m (300–6,500ft)
Parent clouds:	Cumulonimbus or stratocumulus
Rarity:	4/5

Arcus clouds are low horizontal formations that form either in shelf or roll formations.

Shelf cloud

Shelf clouds (opposite top) are long wedge-shaped arcus clouds. They are usually attached to a cumulonimbus cloud but can be formed on cumulus clouds. Strong downdraughts pull cool air from the storm cloud, spreading out under the parent cloud. This cooler air undercuts the lower warmer air, forcing it upwards, condensing it into the visible cloud.

Roll cloud

A roll cloud (opposite bottom) is a detached cloud rotating along a horizontal axis. They are known by the new species name of volutus when there is no other parent cloud present. They can be up to 1,000km long (600 miles) and 2km high (6,500ft), and are usually seen individually, but there can be as many as ten roll clouds in a row. They usually form at the boundaries of a sea breeze, but can also develop from thunderstorm downdraughts.

Did you know?

Roll clouds are most common in Queensland, Australia, where they are known as the Morning Glory cloud and in October form on up to 40 per cent of the days. This is due to the configuration of the land and the sea, although the exact mechanism as to why they form is still unknown.

Kelvin-Helmholtz waves (fluctus)

Height:	Any
Associated clouds:	Cirrus, altocumulus, cumulus, stratus, stratocumulus
Rarity:	5/5

Kelvin-Helmholtz clouds are an extremely rare phenomenon where clouds form in the pattern of breaking waves. They are extremely short-lived, lasting a few seconds or minutes – the only time I spotted this, it had gone in the seconds it took to find my camera!

This breaking wave shape is no coincidence, as it forms in exactly the same way as waves on water. These waves happen when one layer of air flows over another, slower-moving layer of air. In clouds, this is specifically the boundary between a warmer air mass and a colder air mass underneath. This warmer air mass moves faster than the colder air mass, scooping the top of a layer of cloud into waves. Over a short amount of time, the waves build up, curl over and then 'break' and collapse back into the main cloud shape (see diagram).

Sometimes the clouds literally form out of the blue, without forming on top of another cloud. This is because the cooler air is forced to rise, cooling it below dew point and causing clouds as seen in the diagram opposite.

Kelvin-Helmholtz waves

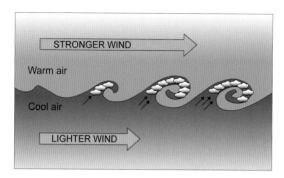

The name Kelvin-Helmholtz comes from the physicists who studied turbulent airflow, Hermann von Helmholtz and Lord Kelvin.

Accessory clouds

Cap cloud (pileus)

Height:	2,000–14,000m (6,500–50,000ft)
Parent clouds:	Cumulus (especially cumulus congestus), cumulonimbus calvus
Rarity:	4/5

Pileus (from the Latin for 'cap') is also known as a scarf or cap cloud. They are small flat clouds that can form just above the tops of cumulus congestus and cumulonimbus calvus clouds. They are formed by a strong updraught forcing moist air above the cloud to rise, where it is cooled below the dew point, causing a layer of small water droplets. The droplets often freeze due to the high altitudes, causing a layer of icy fog. This forms a cap above the top of the cloud as pictured opposite.

When to spot them
The effect is usually very short-lived, as the main cloud continues to rise through convection, absorbing the pileus cap. Sometimes multiple pileus caps are observed on top of each other. Occasionally bright colours are spotted in the cloud, due to iridescence.

Associated weather
The cumulus congestus clouds that bring pileus caps often bring rain. Cumulonimbus calvus clouds bring rain and often hail and thunder. You may escape from the rain if you aren't directly downwind from these clouds. However, just because one cloud missed you doesn't guarantee that there won't be more coming your way!

Pannus

Height:	0–4,000m (0–13,000ft)
Parent clouds:	Nimbostratus, cumulonimbus
Rarity:	2/5

Pannus or 'scud' clouds are ragged or wispy grey clouds that form below the main cloud base. They form due to precipitation falling from above, which moistens the air until it can no longer hold any more water vapour and it condenses into water droplets, forming clouds. They are also known as virga or fall streaks when they are in their wispy form and look similar to the fall streaks from cirrus clouds, but they are technically different as virga is falling precipitation, rather than clouds caused by air moistened by precipitation.

When to spot them
Pannus clouds can form under any raining cloud, so are usually seen under either cumulonimbus or nimbostratus clouds. They themselves are impossible to distinguish from stratus fractus unless the parent cloud is taken into consideration. So if you notice stratus fractus clouds, you should also check for raining clouds above as they may actually be categorised as pannus clouds.

Associated weather
Because they form under rain clouds, Pannus are extremely accurate weather forecasters – after pannus are seen, the rain will reach the ground within ten minutes.

Pannus clouds are fragmented due to the turbulent winds and will move and change shape rapidly. They often increase in number as the rain continues until they merge to become one continuous sheet, obscuring the main cloud base.

Velum

Height:	Any
Parent clouds:	Cumulus (especially cumulus congestus), cumulonimbus
Rarity:	4/5

Velum is a thin veil of cloud that often covers large areas. Its name comes from the Latin for 'sail' or 'veil'. It is linked to cumulus or cumulonimbus clouds, which cause a general uplift of the air around them. Given the right conditions (a stable moist layer), the rising air will condense into a layer similar to thin altostratus. It is generally darker than its parent cumulus or cumulonimbus cloud.

When to spot them

Velum can be caused over the tops of the cumuliform clouds (opposite bottom), or more commonly the cumulus congestus or cumulonimbus cloud breaks through the layer of velum (opposite top) and rises into the clear air above it. Velum can form at any height, but it is important not to get it confused with pileus caps, which are usually very short-lived.

Often, more than one layer of velum is produced as in the top photo opposite. Due to their stable nature, velum can persist for a long time after their parent clouds have decayed.

Associated weather

The weather depends on which type of cumulus or cumulonimbus cloud the velum is associated with (see the section on low clouds for details).

Fog

Fog is not technically a cloud, but like clouds it is made up of water vapour that has condensed into water droplets. Fog's formation is very different to that of clouds and is mostly caused by a cooling of warmer air by the sea (advection fog) or ground (radiation fog). We will also look at upslope fog, frontal fog and sea smoke, which are caused by completely different mechanisms.

Sea fog (advection fog)

Sea fog is known as advection fog and is caused by wind carrying warm humid air over a colder sea. The warm air is cooled by the sea below it, and this cooling means the air cannot hold as much water vapour, so some of it condenses into visible water droplets. Advection fog can be blown into beaches or harbours but is unlikely to make it more than a kilometre or two inland before it dissipates.

Fog banks, like the one in the photo opposite, are advection fog. They can be extremely thick and cover hundreds of miles of sea.

When will it clear?

It is difficult to forecast when advection fog will clear as it doesn't disperse over the course of the day like radiation fog, and can persist for days. A wind shift will usually clear advection fog as the new wind will have come from a different area, so is unlikely to have the same temperature and humidity. An increase in the wind speed mixes the air, so the fog will probably rise and become low stratus cloud if the wind increases above around 10 knots. Additionally, warmer waters may have less fog, so that moving inshore in the summer can increase visibility.

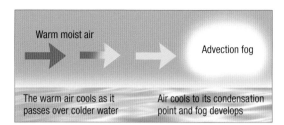

Warm moist air

Advection fog

The warm air cools as it passes over colder water

Air cools to its condensation point and fog develops

When to spot it

The sea is relatively colder inshore in the winter and spring and offshore in the summer and autumn, so fog is more likely inshore in the winter and spring and offshore in the summer and autumn. Overall, in European waters, it is most common in the second half of the year.

Advection fog is often found where cold water currents meet a warmer wind, so places such as eastern Canada and the USA often experience fog, with certain places in Newfoundland experiencing over 200 foggy days per year.

Land fog (radiation fog)

Land fog is caused when the ground cools rapidly (primarily through radiation, hence the technical name, radiation fog). This lowers the air temperature directly above the ground until it reaches dew point and the water vapour in the air condenses into water droplets, which is the fog. Fog forms in a thin layer close to the ground that is no more than tens of metres thick, often meaning that low hills or tops of buildings will be above the fog.

Where to spot it

Radiation fog is most likely to form when the air is moist and the sky is clear, allowing for heat to radiate away easily. It is also more prevalent in winter as the nights are longer so there is more time for the fog to develop. It must be calm, because even a small amount of wind will cause turbulence, mixing the air and dissipating the fog.

Land fog is usually seen over flat land or in valleys. It can also be seen in harbours, where it sinks down the valley sides over the water. It is not seen over open water.

Radiational cooling at the top now deepens the fog

The air is then cooled below saturation point and fog forms

After sunset the ground radiates out heat, cooling the air at the bottom

When will it clear?

The main way it will clear is that the sun will heat up the land below the fog, warming the air back above the dew point. Another way the fog will break up and clear is if the winds increase above about 5 knots, or it starts to rain.

Upslope fog

Upslope fog, as I'm sure you'll have guessed, is fog that is formed up a slope. Moist air is lifted as it flows up a slope, which cools it until it reaches the dew point. Once it's reached the dew point it forms fog, which looks similar to stratus cloud (often specifically stratus fractus) when seen from above or below.

Unlike radiation fog, upslope fog can sustain itself in higher wind speeds (although winds over about 12 knots often disperse it into stratus clouds instead). A slow and steady wind can feed the fog until it becomes even thicker and longer-lasting. Upslope usually forms first high up in the mountain, and slowly thickens until it almost reaches the valley floor.

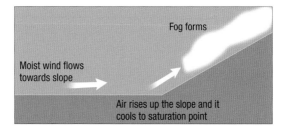

Fog forms

Moist wind flows
towards slope

Air rises up the slope and it
cools to saturation point

When to spot it

Upslope fog is particularly common on sunny days. This is because the sun heats the mountain air quicker than the air in the foothills, causing lower pressure in the

Did you know?

Upslope fog is present for over 80 per cent of the year on Mount Emei in China, as the damp monsoon air is often forced up the mountainside.

mountains than the valleys. Since air flows from areas of higher pressure to lower pressure, the air from the valleys flows up the mountain slopes, causing upslope fog. It needs particularly humid days to form. It will only form on the windward (upwind) side of mountains as the air has to be rising.

When will it clear?

It will usually clear if the wind direction or air mass changes, causing a different temperature or humidity of the air.

Frontal fog

Frontal fog is often neglected as a weather condition, but it is very common in temperate regions like the UK. Again, the name is a bit of a giveaway as it forms in frontal zones. There are actually three types of frontal fog: pre-frontal warm front fog, post-frontal cold front fog and frontal passage fog.

Pre-frontal and post-frontal fog forms when rain falls into colder stable air, evaporating and causing it to become more saturated. Once the cooler air is saturated beyond the dew point, frontal fog appears.

When to spot it

Pre-frontal warm front fog is the most common type and often appears over large areas before warm fronts. Sometimes fog isn't formed, but low stratus forms instead (often from the amalgamation of pannus clouds). This fog normally dissipates once the front has passed by as the temperature increases and the wind changes in direction and strength.

Post-frontal cold front fog is less common and occurs in shallow sloping, slow cold fronts. These give persistent precipitation and can cause fog for up to 400km (250 miles) behind the front. Faster-moving cold fronts don't usually lead to fog, but often produce stratus clouds.

Frontal passage fog occurs as a warm or cold front passes overhead, but clears swiftly afterwards. It forms in a number of ways, but most commonly through the cooling of warm moist air when it mixes with cool moist air.

Sea smoke (steam fog)

Sea smoke, also known as steam fog, is a rare type of fog formed when cold air moves over warmer water. The warmer water both increases the moisture content and warms up the air immediately above it. Cold air then moves over the shallow layer of warm air, cooling it below dew point, when fog is seen. It happens in exactly the same way as the 'steam' from a hot drink is formed.

Sea smoke requires a very light wind as too much wind would completely break up the warmer air layer. Sea smoke itself never rises particularly high, usually only a few metres above the water, so it isn't generally a problem for navigation, except for the smallest craft. It varies from a few thin turbulent shreds of smoke like mist to a dense, but still shallow fog, like that in the photo opposite.

When to spot it

Sea smoke usually forms in polar regions, where cold air from arctic ice flows over a relatively warmer (but still very cold) body of water. In polar regions it is known as Arctic sea smoke. It is most commonly seen in coastal seas around cold land masses, such as Greenland, Norway and Svalbard. It is rarely seen in temperate regions, although it may be seen on cold winter days over rivers and lakes when the air is more than 10°C colder than the water. Opposite is an example of this over Lake Michigan on a freezing winter's day, where the air is considerably colder than the lake.

Turbulence cools warmer air below dew point, causing fog

LIGHT COLD WIND (-10°C)

Frozen land

Warmer sea (1°C)

Clouds outside the troposphere

All the clouds discussed so far are found in the troposphere. However, the next two clouds are much higher, so are not found in the troposphere. Being outside of the troposphere means that they are too remote to have a direct influence on the weather. They are only spotted at night as the sky is too bright to see them during the daytime.

Nacreous (mother of pearl)

Height:	15–25km (9–16 miles)
Rarity:	5/5

Nacreous clouds (technically known as polar stratospheric clouds) are extremely brightly coloured clouds. They are named after the old English word nacre, which means mother of pearl, due to the bright colours.

They are one of two clouds that don't form in the troposphere, instead forming in the stratosphere about twice as high as the highest clouds in the troposphere. Nacreous clouds require extraordinarily cold temperatures of -78ºC (-108°F). They are made up of much smaller particles than in other clouds, causing

Did you know?

These beautiful clouds hide a dark secret. The cloud particles assist the production of chlorine atoms; each atom is estimated to destroy 100,000 ozone molecules, so these clouds are actually the most damaging cloud for the environment.

luminescent colours through iridescence. The clouds are seen after sunset and before sunrise, most often when the sun is between 1 and 6 degrees below the horizon.

When to spot it
These are seen during the winter in the higher latitudes, most commonly over Antarctica, but also in the Arctic, Scotland, Scandinavia, Canada, northern Russia and occasionally in other parts of northern Europe. These are one of the rarest clouds (and are becoming rarer now that CFC gases, which aid the formation of these clouds, aren't being released in the same quantities).

Noctilucent

Height:	75–85km (47–53 miles)
Rarity:	5/5

Noctilucent clouds are the highest clouds, about eight times higher than the 'high' clouds from the troposphere! At about 80km (50 miles) high, they are the only clouds to reside in the mesosphere and are extremely rare. They look like thin milky waves of cirrus clouds and are only seen on clear summer nights, where they shine quite brightly, hence the name noctilucent, meaning night-shining. They are made up of ice crystals that are 1/2500th of the diameter of ice particles from cirrus clouds.

When to spot them

They are only seen on clear summer nights as they are not bright enough to be seen in the day. They are seen when the sun has set well below the horizon but still illuminates these noctilucent clouds from below due to their remarkable height. They are almost always seen between 45 and 65 degrees latitude as it remains astronomical twilight throughout summer nights, allowing the sun to shine onto the clouds.

Scientists still don't know how these clouds were formed. However, it does seem that they are becoming more frequent, but whether this is because of human activity or simply because more people are looking out for these clouds is still up for debate.

Did you know?

The mesosphere is incredibly dry, being one hundred millionth as moist as the air from the Sahara, so noctilucent clouds can only form in intensely cold conditions of less than -120°C (-184°F).

Optical phenomena

Clouds cause many optical phenomena or effects, often giving spectacular colours, or strange brighter patches in the sky. This section will explore what the effects look like, when to spot them and what causes them.

Rainbow

Associated clouds:	Rain clouds (nimbostratus, cumulonimbus)
Rarity:	2/5

A rainbow is an optical illusion that looks like a multicoloured arc. It is always formed at exactly 42 degrees from the antisolar point (the point directly opposite the sun). Rainbows can be seen in any form of water droplets, including sea spray, fog, waterfall spray or even spray from a hosepipe on a sunny day!

A rainbow isn't a physical thing that exists at a specific point in the sky as it depends on the position of the sun and where you as an observer are standing. If you see someone 'at the end' of a rainbow, they will be seeing another rainbow beyond them, similar to yours. Rainbows are usually formed on showery days (below cumulonimbus clouds), as breaks in the cloud are needed for the sunlight to shine through.

Did you know?

Although rainbows are often depicted as seven different distinct colours, it is actually a continuous spectrum of light, with every colour between violet and red. It even contains invisible infrared and ultraviolet light on either side of the bow.

Rainbows are formed by a mix of refraction and reflection. Sunlight entering a water droplet gets refracted (bent). The sunlight is then reflected inside the water droplet and then refracts again as it leaves the droplet. This means that the sunlight is reflected back towards the observer, but is also separated into a continuous spectrum of different colours. Red light always appears on the outside of the arch and violet always appears on the inside.

Double rainbows are sometimes seen, the second rainbow caused by light reflecting twice inside the water droplet. The second rainbow is always above the main rainbow and is much fainter. Its colours are also reversed, with red on the inside and violet on the outside.

Rainbows are actually full circles, but the other half is blocked by the horizon. Circular rainbows are spotted by pilots, or occasionally by people on the top of steep mountains.

Corona

Associated clouds:	Altostratus, altocumulus
Also look for:	Iridescence
Rarity:	2/5

A corona is produced by the diffraction of either moonlight or sunlight. It is a central bright disc with coloured rings around it. In the middle of the corona is the sun or moon. It is most commonly seen around the moon as the sun is usually too bright to see it, but try covering the sun with a hand in the right conditions and you may well spot a corona (just be careful not to look directly at the sun!).

Coronae need thin clouds with similarly sized droplets. Newly forming altostratus clouds are a good example of this. As clouds develop, the droplets start to become irregularly sized, reducing the brightness of the colours. Coronae are usually about 15 degrees in diameter but can increase and decrease in size as the cloud passes through. The size depends on the size of the water droplets, with smaller droplets resulting in larger rings.

Coronae have an extremely bright central disc, which is completely white, with reddish edges. Attached to the disc are rings that start as blues, becoming reds on the outside. This sequence occasionally repeats itself when the conditions are perfect, as in the photo opposite.

Don't mix these up with a halo, which forms a larger ring around the sun or moon.

Halo

Associated clouds:	Cirrostratus, cirrus
Also look for:	Sun dogs
Rarity:	1/5

Haloes often form around the sun or moon, usually due to the presence of cirrostratus. They form because sunlight is reflected by the ice crystals in the sky, forming a white halo. The sky inside of the halo is darker than the sky outside of it as the light is reflected away from this region. Sometimes, as in the example opposite, a faint colouration is observed due to refraction in the ice crystals.

The most frequent halo has a 22-degree radius, so they are called 22-degree haloes. Variations in cloud thickness often mean that the halo is incomplete, forming an arc. A 46-degree halo can also form, but this is a bit rarer and usually only appears as an arc.

Haloes are estimated to be present once every three days but are rarely observed as the brightness of the sun drowns out the halo. However, if you cover the sun with a hand you may well see a halo (being careful not to look directly at the sun).

Haloes can even form around things like street lights in extremely cold conditions, where tiny ice crystals known as 'diamond dust' are held in suspension, reflecting light into a halo.

Associated weather

Depressions are often coming, bringing rain and strong winds. This is because cirrostratus clouds are often seen before the warm or occluded front of a depression arrives.

Mock sun / sun dog / parhelion

Associated clouds:	Any clouds made from ice crystals, especially cirrus and cirrostratus
Also look for:	Halo
Rarity:	3/5

A parhelion (from the Greek for 'with the sun') is a brighter patch of light observed to one or both sides of the sun. They are always at the same height as the sun. They can either be white patches of light or, more commonly, have a faint rainbow effect, like the photo opposite.

They are formed by refraction from horizontally aligned ice crystals and are usually seen on the outside of the 22-degree halo. The red end of the rainbow spectrum is always on the inside of the sun dog. They are most obvious when the sun is low in the sky and can't be seen if the sun is higher than 60 degrees.

Moon dogs are also possible, but are much rarer and are technically called a paraselene.

Iridescence

Associated clouds:	Altocumulus, altostratus, cirrocumulus, cirrus and nacreous
Also look for:	Corona
Rarity:	2/5

Iridescence (also known as irisation) is a colourful optical phenomenon. It can appear as bright rainbow colours, but is normally a pastel colour. It is usually spotted on the edges of clouds, but can be seen in the middle of thin clouds.

It can be spotted briefly in almost any cloud, but is most common in altocumulus, altostratus, cirrocumulus and cirrus clouds. Iridescence is particularly vivid in nacreous clouds, but is not often seen as nacreous clouds are extremely rare. Iridescence is a very common phenomenon but is often not noticed as it forms close to the sun, so it gets drowned out unless the sun is covered with a hand or another object. It is most striking at about 30 degrees from the sun.

It is formed due to sunlight or moonlight diffracting through the small ice particles or water droplets. The particles must be small, otherwise haloes will be caused by refraction instead. The cloud also needs to be thin enough to allow each sunray to meet only one particle, otherwise the light is scattered in too many directions. The most vivid colours are seen when the particles are all a very similar size, whereas pastel colours are seen when there is more variation in particle size.

Sometimes a corona is formed instead. A corona would form when the particles in the sky are similar in size and also cover a large area in a uniform layer. Coronae are coloured rings connected to a central bright disc around the sun.

Ice rainbow (circumzenithal arc)

Associated clouds:	Cirrostratus, cirrus
Also look for:	Halo
Rarity:	3/5

A circumzenithal arc (also known as an upside-down rainbow) is the optical phenomenon that has the most vibrant spectrum. It is formed when sunlight refracts through horizontally aligned ice crystals (just like the crystals in sun dogs). For this reason, it is usually seen in frozen cirrostratus clouds (opposite top). It can also be seen in cirrus clouds if they cover significant areas of the sky, but this is less common.

It is part of a circle (usually a quarter of a circle) that is centred around the zenith (the point immediately above you). It can't occur if the sun is higher than 32 degrees and is brightest when the sun is 22 degrees above the horizon.

It has a spectacular spectrum of colours, with blue nearest the zenith and red nearest the horizon. It shouldn't be confused with a halo, which is much less vibrant than a circumzenithal arc, and curves around the sun, whereas the circumzenithal arc is convex to the sun (it curves away from the sun).

A rarer form is a circumhorizontal arc (opposite bottom) which is a band of colour parallel to the horizon. This is only seen when the sun is high in the sky (above 58 degrees), so is only occasionally seen in midsummer in temperate regions. It can be confused with a rainbow, but the colours are reversed, with red being closest to the sun instead of blue.

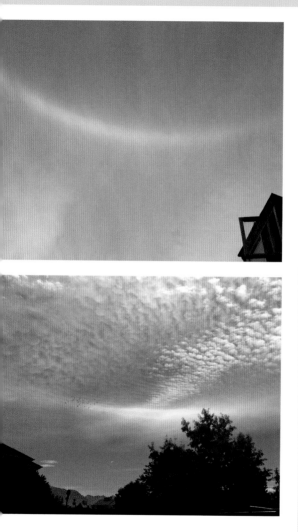

Sunbeams and crepuscular rays

Associated clouds:	Altocumulus, stratocumulus
Rarity:	1/5

Rays of sunlight are completely invisible, so how is it sometimes we seem to see sunrays coming out between clouds? This is because the sunlight bounces off particles in the atmosphere. These are not water molecules, but dry particles such as dust or smoke, scattering the light back to us in spectacular displays. This means that sunbeams are much less likely to be seen immediately after rain, where the droplets have 'washed' the dust out of the sky.

Looking at the photos opposite you would be excused for thinking that the light rays come out in different directions, but they are actually parallel as they are all coming from the same light source. It is just perspective that makes sunbeams appear to diverge.

Crepuscular rays are sunbeams originating after sunset, such as in the photo opposite, top. These often have beautiful yellow or red colours as the sunset light has to pass through up to 40 times more air than it would at midday. This means

that more of the blue and green light is scattered by the atmosphere than the yellow and red light.

On occasion, crepuscular rays extend across the whole sky and converge on the opposite horizon; in which case they are called anticrepuscular rays or antisolar rays (see photo below).

Glory

Associated clouds:	Stratus, upslope fog
Rarity:	3/5

A glory is an optical phenomenon made up of coloured concentric rings at the antisolar point (the point directly opposite the sun). It is seen in clouds of similarly sized water droplets. It is caused by light refracting within water droplets, which is then scattered back towards the sun, interfering with the other light waves, causing a coloured spectrum. The colours are red on the outside and blue towards the centre, and can repeat themselves, steadily becoming dimmer further from the centre. It is not to be confused with a circular rainbow.

A Brocken spectre is seen when the shadow of an observer partially obscures the glory. This gives the effect of a magnified shadow, with rings around the observer's head. This gives a ghost-like effect, as in the photo opposite. It was first recorded on the Brocken mountain in Germany, hence the name. The shadow seems magnified as it is actually cast on nearby cloud, but appears to be cast on the landscape in the far distance.

Did you know?

Glories are seen from mountains, tall buildings, or most commonly aeroplanes. This is because they have to form below the observer as they occur directly opposite the sun, which is always above an observer.

Cloud weather lore

I am sure we all know some weather lore, but is any of it actually useful to us now? In this section, you will find out why some of these ancient proverbs work and why others aren't nearly so reliable.

Red sky at night, shepherd's delight
Red sky in the morning, shepherd's warning

Not only is this the most well-known weather proverb, it is also one of the oldest recorded weather sayings. Two thousand years ago, Matthew's Gospel in the Bible recorded Jesus saying: *When evening comes, you say, 'The weather will be fair, for the sky is red', and in the morning, 'Today it will be stormy, for the sky is red and overcast.'*

This proverb is generally true due to the position of the clouds and the sun. For a red sky to occur, it must be cloudy on the side opposite the setting or rising sun and clear on the sun's side, allowing the sun to illuminate the clouds. If the sky is red in the evening, it means that there must be clear weather in the west, but cloud in the east. Since weather systems generally move from west to east, this means that the poor weather has already passed through, so there should be good weather the next day. The opposite is true with a red sky in the morning, where there is poor cloudy weather to the west and clear skies to the east, so a poor weather system such as a warm front may be coming.

This weather lore is only correct in the mid-latitudes of both hemispheres, as that is where weather systems travel from west to east. It is also very common that gaps in clouds can form, causing a red sky completely coincidentally, rather than being linked to an incoming or outgoing weather system.

Cloud weather lore

Mares' tails and mackerel scales make tall ships carry low sails

'Mares' tails' refers to cirrus uncinus clouds (page 16) and mackerel scales refer to cirrocumulus or altocumulus clouds. These skies are often seen before a depression comes as they can be the first indication of a warm front. This generally means that rain will start in 12–24 hours' time and the winds will increase dramatically as the depression passes. This indicates to ships that they need to reef their sails, decreasing their size in time for the storm. This can be quite accurate as cirrus clouds are almost always seen before a depression, but they can also be seen at other times. Mackerel skies are often seen before a depression, but again are often seen at other times.

Trace in the sky the painter's brush, the winds around you soon will rush

Similar to the previous one, the painter's brush refers to the wispy cirrus clouds that look like an artist's brushstroke. Since cirrus clouds are often seen before depressions, they are considered indicators of an increasing wind.

If it rains before seven, it will clear before eleven

This is based on the fact that weather systems usually pass through quite quickly, often not giving rain for more than four hours in a row. This can be true if there is a single warm front passing through. However, if the front is slow-moving or there are multiple fronts, we often have a full day of rain – as I'm sure you've often experienced at the wrong times! So I certainly wouldn't rely on this, but it can be true.

Cloud weather lore

When clouds appear like rocks and towers, the earth's refreshed with frequent showers

This one refers to towering cumulus congestus and cumulonimbus clouds. Cumulonimbus calvus and cumulus congestus clouds tower kilometres high and their bulging appearances look a bit like rocks. Cumulonimbus clouds bring heavy showers and even thunder and lightning, so seek shelter quickly if you see these clouds upwind of you!

Mountains in the morning, fountains in the evening

Tall cumulus clouds (cumulus congestus) tower up like mountains in the sky. If these are seen in the morning, it means that the air is humid and they are likely to keep growing in size. As they develop further they grow into the taller cumulonimbus clouds, which bring heavy showers in the afternoon. By the evening the sun loses a lot of its heat, so the cumulonimbus clouds stop growing and start to decay, making rain less likely. A more accurate version of this would be 'mountains in the morning, fountains in the afternoon' – it just doesn't have the same ring to it!

The more cloud types at dawn, the greater the chance of rain

This is often true, assuming that the cloud types are high. This is because lots of different cloud types, such as cirrus, cirrostratus and cirrocumulus, are seen before a depression comes, which brings rain. Most other cloud types are usually seen with only one type present at each time, so it would be more accurate to say, 'the more high cloud types…'

If cirrus clouds dissolve and appear to vanish, it is an indication of fine weather

Cirrus clouds that don't thicken, but appear and disappear, don't indicate an incoming depression. Instead, they mean the weather should stay fair.

When clouds look like black smoke, a wise man will put on his cloak

This basically is saying that dark clouds bring rain, which is true to a degree. Thicker clouds block more sunlight, so are darker. Rain clouds need to be dense and thick to rain, so are generally darker. You should look at the cloud types themselves to see if they may bring rain, or only look dark because they are shaded by other clouds.

Glossary

Air mass: Body of air with constant temperature and humidity.

Air pressure: The weight of the air above a particular point.

Anticyclone: Area of higher air pressure.

Cold front: Where a cold air mass undercuts a warmer air mass, bringing heavy showers.

Depression: Area of lower air pressure.

Dew point: The temperature below which the air becomes fully saturated and the water vapour condenses into water droplets.

Diffraction: Where waves spread out when passing through a gap. Different colours spread out by different amounts, separating them into the spectrum of colour that we see.

Fully saturated: The humidity at which the air can hold no more water vapour without it condensing. It is known as 100 per cent humidity.

Front: A narrow zone that separates cold and warm air masses.

Occluded front: A front where the cold front has caught up with the warm front, forcing the warm sector above the ground.

Refraction: Where waves change direction when passing from one medium to another. For example, sunlight passing from air to ice crystals or water droplets, causing iridescence, or sunlight at sunset, causing coloured sky.

Temperature inversion: A layer in the atmosphere where the air temperature increases with height.

Tropics: The region between the tropic of Cancer (23.5 degrees north) and the tropic of Capricorn (23.5 degrees south).

Tropopause: The top boundary of the troposphere. It is a temperature inversion, so the temperature increases with height after this.

Troposphere: The lowest layer of the atmosphere, which varies between 6km (4 miles) thick in polar regions and 18km (11 miles) thick over the Equator.

Warm front: Where a warm air mass rises over a cooler, denser air mass. This brings steadily thickening cloud and then rain, which slowly increases in intensity as the front approaches.

Warm sector: The area of relatively warmer air found between a cold front and a warm front.

Quick cloud identifier

Use this to help identify what cloud you are looking at. It will identify all the main cloud types as well as a couple of common accessory clouds and supplementary features. You should read the cloud's entry to confirm the identification as sometimes more information is needed than this identifier requires.

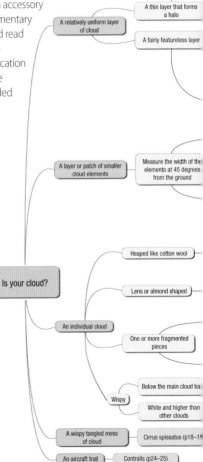

Is your cloud?

A relatively uniform layer of cloud
- A thin layer that forms a halo
- A fairly featureless layer

A layer or patch of smaller cloud elements
- Measure the width of the elements at 45 degrees from the ground

An individual cloud
- Heaped like cotton wool
- Lens or almond shaped
- One or more fragmented pieces

Wispy
- Below the main cloud ba...
- White and higher than other clouds

A wispy tangled mess of cloud — Cirrus spissatus (p18–19...)

An aircraft trail — Contrails (p24–25)

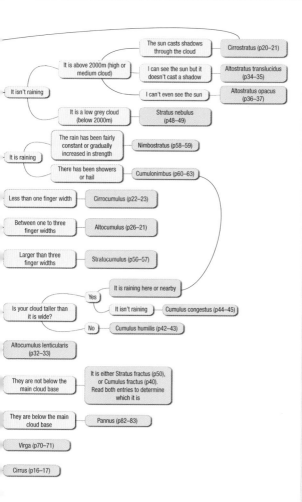

The sun casts shadows through the cloud — Cirrostratus (p20–21)

It is above 2000m (high or medium cloud)

I can see the sun but it doesn't cast a shadow — Altostratus translucidus (p34–35)

I can't even see the sun — Altostratus opacus (p36–37)

It isn't raining

It is a low grey cloud (below 2000m) — Stratus nebulus (p48–49)

The rain has been fairly constant or gradually increased in strength — Nimbostratus (p58–59)

It is raining

There has been showers or hail — Cumulonimbus (p60–63)

Less than one finger width — Cirrocumulus (p22–23)

Between one to three finger widths — Altocumulus (p26–21)

Larger than three finger widths — Stratocumulus (p56–57)

It is raining here or nearby

Yes

It isn't raining — Cumulus congestus (p44–45)

Is your cloud taller than it is wide?

No — Cumulus humilis (p42–43)

Altocumulus lenticularis (p32–33)

They are not below the main cloud base — It is either Stratus fractus (p50), or Cumulus fractus (p40). Read both entries to determine which it is

They are below the main cloud base — Pannus (p82–83)

Virga (p70–71)

Cirrus (p16–17)

125

Index

Note: *italic* page numbers indicate figures.

accessory clouds 66, 80–1, *80–1*
air masses 12–13, 122
air pollution 14, 54
air pressure 12, 13, 70, 122
aircraft *see* contrails
altocumulus 7, *9*, 10, 22, 26–33, 66, 70, 78, 102, 108, 112
altocumulus castellanus 27, 28–9, *29*, 70
altocumulus floccus 26–7, *27*, 70
altocumulus lenticularis 32–3, *33*
altocumulus sheet (stratiformis) 30–1, *30–1*
altostratus 7, *9*, 10, 26, 34–7, 50, 68, 102, 108
altostratus radiatus 34
altostratus, thick (opacus) 34, 36–7, *37*, 58
altostratus, thin (translucidus) 34–5, *35*
anticyclones 14, *15*, 122
anvil (incus) 8, 62, 64, 72–3, *73*
arcus clouds 62, 76–7, *77*
asperitas 8, 54, 66–7, *67*

Brocken spectre 114, *115*

cap cloud *see* pileus
capillatus 61, 62–3, *63*
castellanus 16, 22, 27, 52, 56–7, *57*, 70
cavum 22, 54
circumhorizontal arc 110, *111*
circumzenithal arc 20, 110–11, *111*
cirriform 24

cirrocumulus 7, *9*, 16, 22–3, *23*, 108
cirrostratus 7, *9*, 16, 20–1, *21*, 104, 120
cirrus 7, *9*, 10, 16–17, *17*, 34, 68, 70, 72, 78, 104, 108, 120
cirrus homogenitus 24–5, *25*
cirrus spissatus 18, *19*
cirrus uncinus 16, 17, 118
Cloud Appreciation Society 66
cloud classification 7–9, *9*, 10–11
cloud species/genera 7–8
cloud streets (cumulus radiatus) 46–7, *47*
cold fronts 13, 14, 31, 92, 122
congestus 38, 42, 43
contrails 24–5, *25*
coronas 102–3, *103*, 108
crepuscular rays 54, 112, *113*
cumuliform clouds 84
cumulonimbogenitus 18, 72
cumulonimbus 7, 27, 38, 57, 58, 60–3, 64, 68, 70, 74, 82, 84, 100, 120
   and cirrus spissatus 18
   see also anvil
cumulonimbus calvus 60–1, *61*, 63, 64, 80, 81
cumulonimbus capillatus 61, 62–3, *63*, 64, 72
cumulus 7, *9*, 10, 38–47, *39*, 80, 84
cumulus cloud street (cumulus radiatus) 46–7, *47*
cumulus congestus 42, 43, 44–5, *45*, 60, 70, 74, 80, 81, 84, 120
cumulus fractus 40–1, *41*, 42, 50

cumulus humilis 40, 42–3, *43*
cumulus mediocris 42, 43

depressions 13–14, *14–15*, 35, 104, 120, 122
downdraughts 64, 68, 76
duplicatus 8, 16, 20, 34, 36, 54, 56

fall streaks *see* virga
fibratus 8, 16
flammagenitus 9
floccus 16, 22, 26–7
fluctus 16, 42, 48, 54, 56
    see also Kelvin-Helmholtz waves
fog 6, 86–95
    frontal 86, 92–3, *93*
    land/radiation 86, 88–9, *88–9*
    sea smoke 86, 94–5, *95*
    sea/advection 86–7, *86*, *87*
    upslope 86, 90–1, *90*, *91*
fractus 38
funnel cloud (tuba) 44, 62, 74–5, *75*

glories 114–15, *115*

haloes 102, 104–5, *105*
homogenitus 8–9, 24–5, *25*
homomutatus 24
Howard, Luke 10, *11*
humilis 8, 38

ice rainbow
    see circumzenithal arc
incus *see* anvil
International Cloud Atlas 11, 66
intortus 16
iridescence 108–9, *109*

Kelvin-Helmholtz waves 78–9, *78*, *79*

lacunosus 22, 54
lenticularis 22, 32–3, *33*, 52
lightning *26–7*, 61, 62, 63, 64, 65, 72, 120

mackerel skies 22
mamma/mammatus 16, 22, 54, 62, 68–9, *69*, 72
mares'tails 118, *118–19*
mediocris 38, 42, 43
mesosphere 98, 99
mock suns 18, 20, 106–7, *107*
Morning Glory cloud 77

nacreous clouds 9, 96–7, *96–7*, 108
nebulosus 20, 48–9, *49*
nimbostratus 7, 9, 26, 50, 58–9, *59*, 63, 70, 82, 100
noctilucent clouds 9, 98–9, *98–9*

occluded front 13, 20, 24, 34, 104, 122
opacus 48, 54

pannus 8, 44, 58–9, 62, 63, 82–3, *83*
parhelion 106–7, *107*
perlucidus 31, 54
pileus 8, 44, 62, 80–1, *80–1*
praecipitatio 44, 48, 54, 58, 62, 66, 71
Pretor-Pinney, Gavin 66
proverbial weather sayings 116–21

radiation fog 86, 88–9, *88–9*

Index

radiatus 16, 34, 36, 40, 42, 44, 46–7, 47
rainbows 100 1, 101
Renou, Émilien 10
roll clouds 76, 77

sea smoke 86, 94–5, 95
shelf clouds 76, 77
short-lived phenomena 78–9, 78, 79, 80, 84
spissatus 18, 19
stratiformis 22, 52
stratocumulus 7, 9, 10, 38, 52–7, 66, 70, 78, 112
stratocumulus castellanus 56–7, 57
stratocumulus stratiformis 54–5, 55
stratus 7, 9, 10, 38, 48–51, 78
stratus, broken (stratus fractus) 50–1, 51
stratus, featureless (stratus nebulosus) 48–9, 49
street lights 104
sun dogs 104
sunrays/sunbeams 112–13, 113

temperature inversions 7, 14, 42, 52, 53, 54, 62, 72, 122
thunderstorms 27, 28, 40, 57, 64–5, 65, 68, 81
tornados 65, 74
translucidus 8, 48, 54
troposphere 7, 9, 123
    clouds above 96–9, 96–7, 98–9
tuba see funnel cloud

uncinus 16
undulatus 22, 34, 36
updraughts 38, 42, 46, 64, 65, 80

velum 44, 62, 84–5, 85
vertebratus 16
virga 22, 44, 56, 58, 62, 70–1, 71
volutus 11, 76

warm fronts 13, 14, 17, 18, 20, 22, 23, 30, 92, 116, 118, 123
weather 6–7, 12–15
    lore 116–21
World Meteorological Organization (WMO) 7, 8

Photo credits

Flickr: Rick Bohn / USFWS p106, Noel Feans p79. **Getty Images**: Robert Alexander pp39, 63, 73 top; Chicago Tribune p95; Ian Forsyth p97; Jeff Greenberg p121; Randy Halverson/Barcroft Media p21 top; Derek Hudson p117; George Rose p33; Galen Rowell p115; Science & Society Picture Library/SSPL p11. **Bastian Kirsch** (distributed via imaggeo.egu. eu) p81. **Pixabay**: Mary Berg p75; Tobias Hämmer p77 top; marcelkessler p112; sebastianmichalke p91; stokpic p113 top. **Wikimedia Commons**: Adrian.lifa p109; Bidgee pp29, 85 top; Kevin Cho (Kee Pil Cho) p99; Daniela Mirner Eberl p77 bottom; Simon A. Eugster p71; Frankie Fouganthin p93; John Fowler p113 bottom; GerritR pp17, 19 bottom, 41, 83 top; GRAHAMUK p78; The Great Cloudwatcher p37; Mikell Johnson p111 bottom; Hussein Kefel p73 bottom; King of Hearts p23; Lauri Kosonen p101; Mika-Pekka Markkanen p103; Ave Maria Möistlik p67, NASA p15 top; Nicholas_T p61; James St John p85 bottom; Teabeer p111 top; Anton Yankovyi p105